~A BINGO BOOK~

# Subtraction Bingo Book

## COMPLETE BINGO GAME IN A BOOK

I0071526

**Written By Rebecca Stark**
**Educational Books 'n' Bingo**

TITLE: Subtraction Bingo
AUTHOR: Rebecca Stark

© Barbara M. Peller, also known as Rebecca Stark, 2016

The purchase of this book entitles the buyer to exclusive reproduction rights of the student activity pages for his or her class only. The reproduction of any part of the work for an entire school or school system or for commercial use is prohibited.

ISBN 978-0-87386-430-5

**Educational Books 'n' Bingo**

Printed in the U.S.A.

# SUBTRACTION BINGO DIRECTIONS

**INCLUDED:**

List of Terms

Templates for Additional Terms and Clues

2 Clues per Term

30 Unique Bingo Cards

Markers

1. **Either cut apart the book or make copies of ALL the sheets. You might want to make an extra copy of the clue sheets to use for introduction and review. Keep the sheets in an envelope for easy reuse.**

2. Cut apart the call cards with terms and clues.

3. Pass out one bingo card per student. There are enough for a class of 30.

4. Pass out markers. You may cut apart the markers included in this book or use any other small items of your choice.

5. Decide whether or not you will require the entire card to be filled. Requiring the entire card to be filled provides a better review. However, if you have a short time to fill, you may prefer to have them do the just the border or some other format. Tell the class before you begin what is required.

6. There are 50 terms. Read the list before you begin. If there are any terms that have not been covered in class, you may want to read to the students the term and clues before you begin.

7. There is a blank space in the middle of each card. You can instruct the students to use it as a free space or you can write in answers to cover terms not included. Of course, in this case you would create your own clues. (Templates provided.)

8. Shuffle the cards and place them in a pile. Two or three clues are provided for each term. If you plan to play the game with the same group more than once, you might want to choose a different clue for each game. If not, you may choose to use more than one clue.

9. Be sure to keep the cards you have used for the present game in a separate pile. When a student calls, "Bingo," he or she will have to verify that the correct answers are on his or her card AND that the markers were placed in response to the proper questions. Pull out the cards that are on the student's card keeping them in the order they were used in the game. Read each clue as it was given and ask the student to identify the correct answer from his or her card.

10. If the student has the correct answers on the card AND has shown that they were marked in response to the *correct questions,* then that student is the winner and the game is over. If the student does not have the correct answers on the card OR he or she marked the answers in response to *the wrong questions,* then the game continues until there is a proper winner.

11. If you want to play again, reshuffle the cards and begin again.

## Have fun!

© Barbara M. Peller

# TERMS/ANSWERS

| | |
|---|---|
| = | 24 |
| − | 30 |
| 1/9 | 33 |
| 1/6 | 36 |
| 3/8 | 40 |
| 1 | 42 |
| 2 | 44 |
| 3 | 45 |
| 4 | 50 |
| 5 | 54 |
| 6 | 60 |
| 7 | 70 |
| 8 | 80 |
| 9 | 90 |
| 10 | ADDITION |
| 11 | BORROW |
| 12 | DIFFERENCE |
| 13 | DIGIT |
| 14 | HUNDREDS PLACE |
| 15 | MINUEND |
| 16 | ONES PLACE |
| 17 | SUBTRACTION |
| 18 | SUBTRAHEND |
| 19 | TENS PLACE |
| 20 | ZERO |

© Barbara M. Peller

# Additional Terms

Choose as many subtraction terms as you would like and write them in the squares. Repeat each as desired.
Cut out the squares and randomly distribute them to the class.
Instruct the students to place their square on the center space of their card.

|  |  |  |  |  |
|---|---|---|---|---|
|  |  |  |  |  |
|  |  |  |  |  |
|  |  |  |  |  |
|  |  |  |  |  |
|  |  |  |  |  |
|  |  |  |  |  |

© Barbara M. Peller

# Clues for Additional Terms

Write three clues for each of your subtraction terms.

| | |
|---|---|
| _____<br><br>1.<br><br>2.<br><br>3. | _____<br><br>1.<br><br>2.<br><br>3. |
| _____<br><br>1.<br><br>2.<br><br>3. | _____<br><br>1.<br><br>2.<br><br>3. |
| _____<br><br>1.<br><br>2.<br><br>3. | _____<br><br>1.<br><br>2.<br><br>3. |

© Barbara M. Peller

| + − | + − | + − | + − | + − |
|-----|-----|-----|-----|-----|
| X ÷ | X ÷ | X ÷ | X ÷ | X ÷ |
| + − | + − | + − | + − | + − |
| X ÷ | X ÷ | X ÷ | X ÷ | X ÷ |
| + − | + − | + − | + − | + − |
| X ÷ | X ÷ | X ÷ | X ÷ | X ÷ |
| + − | + − | + − | + − | + − |
| X ÷ | X ÷ | X ÷ | X ÷ | X ÷ |
| + − | + − | + − | + − | + − |
| X ÷ | X ÷ | X ÷ | X ÷ | X ÷ |
| + − | + − | + − | + − | + − |
| X ÷ | X ÷ | X ÷ | X ÷ | X ÷ |
| + − | + − | + − | + − | + − |
| X ÷ | X ÷ | X ÷ | X ÷ | X ÷ |

| | |
|---|---|
| **=** <br> 1. This sign means "equals." <br> 2. Whatever is left of this sign has the same value as what is to the right of this sign. <br> 3. Fill in the sign: 2 – 2 ____ 4. | **–** <br> 1. This sign means "minus." <br> 2. Fill in the sign: 5 ____ 2 = 3. <br> 3. The process of subtraction is denoted by this sign. |
| **1/9** <br> 1. 8/9 – 7/9 = ____ <br> 2. 6/9 – 5/9 = ____ <br> 3. 5/9 – 4/9 = ____ | **1/6** <br> 1. 5/6 – 4/6 = ____ <br> 2. 4/6 – 3/6 = ____ <br> 3. 3/6 – 2/6 = ____ |
| **3/8** <br> 1. 7/8 – 4/8 = ____ <br> 2. 6/8 – 3/8 = ____ <br> 3. 5/8 – 2/8 = ____ | **1** <br> 1. 8 – 7 = ____ <br> 2. 15 – 14 = ____ <br> 3. 21 – 20 = ____ |
| **2** <br> 1. 8 – 6 = ____ <br> 2. 15 – 13 = ____ <br> 3. 20 – 18 = ____ | **3** <br> 1. 9 – 6 = ____ <br> 2. 10 – 7 = ____ <br> 3. 12 – 9 = ____ |
| **4** <br> 1. 9 – 5 = ____ <br> 2. 8 – 4 = ____ <br> 3. 11 – 7 = ____ | **5** <br> 1. 10 – 5 = ____ <br> 2. 14 – 9 = ____ <br> 3. 13 – 8 = ____ |

Subtraction Bingo

© **Barbara M. Peller**

**6**

1. 12 − 6 = ___
2. 14 − 8 = ___
3. 11 − 5 = ___

**7**

1. 16 − 9 = ___
2. 14 − 7 = ___
3. 15 − 8 = ___

**8**

1. 17 − 9 = ___
2. 18 − 10 = ___
3. 16 − 8 = ___

**9**

1. 12 − 3 = ___
2. 18 − 9 = ___
3. 17 − 8 = ___

**10**

1. 20 − 10 = ___
2. 30 − 20 = ___
3. 12 − 2 = ___

**11**

1. 13 − 2 = ___
2. 15 − 4 = ___
3. 33 − 22 = ___

**12**

1. 24 − 12 = ___
2. 36 − 24 = ___
3. 48 − 36 = ___

**13**

1. 26 − 13 = ___
2. 15 − 2 = ___
3. 19 − 6 = ___

**14**

1. 18 − 4 = ___
2. 28 − 14 = ___
3. 26 − 12 = ___

**15**

1. 20 − 5 = ___
2. 30 − 15 = ___
3. 19 − 4 = ___

Subtraction Bingo

© Barbara M. Peller

| | |
|---|---|
| **16**<br>1. $19 - 3 =$ \_\_\_<br>2. $32 - 16 =$ \_\_\_<br>3. $20 - 4 =$ \_\_\_ | **17**<br>1. $24 - 7 =$ \_\_\_<br>2. $34 - 17 =$ \_\_\_<br>3. $21 - 4 =$ \_\_\_ |
| **18**<br>1. $24 - 6 =$ \_\_\_<br>2. $36 - 18 =$ \_\_\_<br>3. $20 - 2 =$ \_\_\_ | **19**<br>1. $21 - 2 =$ \_\_\_<br>2. $24 - 5 =$ \_\_\_<br>3. $22 - 3 =$ \_\_\_ |
| **20**<br>1. $30 - 10 =$ \_\_\_<br>2. $28 - 8 =$ \_\_\_<br>3. $22 - 2 =$ \_\_\_ | **24**<br>1. $34 - 10 =$ \_\_\_<br>2. $56 - 32 =$ \_\_\_<br>3. $48 - 24 =$ \_\_\_ |
| **30**<br>1. $40 - 10 =$ \_\_\_<br>2. $32 - 2 =$ \_\_\_<br>3. $36 - 6 =$ \_\_\_ | **33**<br>1. $40 - 7 =$ \_\_\_<br>2. $35 - 2 =$ \_\_\_<br>3. $54 - 21 =$ \_\_\_ |
| **36**<br>1. $40 - 4 =$ \_\_\_<br>2. $38 - 2 =$ \_\_\_<br>3. $37 - 1 =$ \_\_\_ | **40**<br>1. $50 - 10 =$ \_\_\_<br>2. $45 - 5 =$ \_\_\_<br>3. $42 - 2 =$ \_\_\_ |

Subtraction Bingo

© Barbara M. Peller

**42**

1. 46 − 4 = ___
2. 52 − 10 = ___
3. 48 − 6 = ___

**44**

1. 55 − 11 = ___
2. 48 − 4 = ___
3. 46 − 2 = ___

**45**

1. 48 − 3 = ___
2. 45 − 0 = ___
3. 55 − 10 = ___

**50**

1. 52 − 2 = ___
2. 55 − 5 = ___
3. 60 − 10 = ___

**54**

1. 56 − 2 = ___
2. 57 − 3 = ___
3. 64 − 10 = ___

**60**

1. 70 − 10 = ___
2. 65 − 5 = ___
3. 60 − 0 = ___

**70**

1. 80 − 10 = ___
2. 72 − 2 = ___
3. 75 − 5 = ___

**80**

1. 90 − 10 = ___
2. 85 − 5 = ___
3. 100 − 20 = ___

**90**

1. 100 − 10 = ___
2. 95 − 5 = ___
3. 92 − 2 = ___

Subtraction Bingo

**Addition**

1. ___ is opposite that of subtraction.
2. To check a subtraction problem use ___.
3. ___ and subtraction are inverse operations. That means that if 11 − 3 = 8, then 8 + 3 = 11.

© Barbara M. Peller

## Borrow

1. To solve 17 − 8, we have to do this.

2. To solve 32 − 9, we have to do this.

3. To solve 12 − 9, we have to ___ a set of tens and add it to the ones column.

## Difference

1. The result of subtracting two numbers.

2. In the problem 13 − 4, the ___ is 9.

3. The answer in a subtraction problem.

## Digit

1. Any of the numerals 1 to 9 and 0.

2. 98 is a 2-___ number.

3. 653 is a 3-___ number.

## Hundreds Place

1. The place three to the left of the decimal point.

2. In the number 298, the 2 is in this place.

3. In the number 865, the 8 is in this place.

## Minuend

1. In subtraction, the number which is decreased.

2. In the problem 10 − 6 = 5, the 10 is this.

3. In the problem 17 − 9 = 8, the 17 is this.

## Ones Place

1. The place just to the left of the decimal point.

2. In the number 17, the 7 is in this place.

3. In the number 215, the 5 is in this place.

## Subtraction

1. One of the 4 basic arithmetic operations. The others are addition, multiplication and division.

2. Its inverse operation is addition.

3. The process of finding the difference between two numbers or quantities.

## Subtrahend

1. It is the quantity or number to be subtracted from another.

2. In the problem 10 − 6 = 5, the 6 is this.

3. In the problem 17 − 9 = 8, the 9 is this.

## Tens Place

1. The place two to the left of the decimal point.

2. In the number 849, the 4 is in this place.

3. In the number 572, the 7 is in this place.

## Zero

1. It means "none" and is neither positive nor negative.

2. The difference of any number minus this is that number.

3. 12 − ___ = 12

© Barbara M. Peller

# Subtraction Bingo

| 30 | 24 | 3/8 | 13 | 9 |
|---|---|---|---|---|
| 6 | – | Zero | Borrow | 36 |
| 1/6 | Hundreds Place | | 80 | 54 |
| 1/9 | Minuend | 18 | 44 | 70 |
| 90 | 10 | 7 | 15 | 20 |

Subtraction Bingo: Card No. 1

© Barbara M. Peller

# Subtraction Bingo

| 1/9 | 1/6 | 40 | Difference | Subtraction |
|---|---|---|---|---|
| 70 | Borrow | 2 | Minuend | 45 |
| Addition | 10 | | 8 | 18 |
| 14 | 17 | Hundreds Place | 16 | 36 |
| 20 | Zero | 7 | 6 | 15 |

Subtraction Bingo: Card No. 2

© Barbara M. Peller

# Subtraction Bingo

| 1/9 | 18 | Borrow | 44 | 1/6 |
|---|---|---|---|---|
| 10 | – | 3 | 24 | 33 |
| Minuend | Zero | | 45 | = |
| Hundreds Place | Addition | 90 | 14 | 40 |
| 15 | 6 | 7 | 16 | Subtraction |

© Barbara M. Peller

# Subtraction Bingo

| | | | | |
|---|---|---|---|---|
| Hundreds Place | 45 | 3/8 | 6 | Subtraction |
| 42 | 1 | 24 | Difference | 1/6 |
| 80 | 14 | | 9 | 13 |
| 18 | Digit | Zero | 7 | 2 |
| 19 | 20 | Subtrahend | 15 | 54 |

Subtraction Bingo: Card No. 4

© Barbara M. Peller

# Subtraction Bingo

| | | | | |
|---|---|---|---|---|
| 20 | 9 | Minuend | 2 | 6 |
| 42 | 18 | 3 | 8 | – |
| 3/8 | 54 | | 50 | 11 |
| 36 | Subtraction | 30 | 16 | 19 |
| Borrow | 7 | 1/6 | Hundreds Place | 80 |

Subtraction Bingo: Card No. 5

© Barbara M. Peller

# Subtraction Bingo

| = | 45 | 40 | Subtraction | 54 |
|---|----|----|-------------|-----|
| 44 | Minuend | 19 | 24 | 1/6 |
| Difference | 4 | | 1 | 8 |
| 7 | 90 | 16 | Subtrahend | 3/8 |
| 70 | 18 | 30 | 80 | Digit |

© Barbara M. Peller

# Subtraction Bingo

| | | | | |
|---|---|---|---|---|
| 30 | 45 | 11 | 50 | Borrow |
| 70 | Subtraction | 10 | – | 42 |
| 40 | 13 | | 8 | 1 |
| Hundreds Place | 14 | 3 | 1/9 | Addition |
| 7 | 6 | 16 | Subtrahend | = |

Subtraction Bingo: Card No. 7

© Barbara M. Peller

# Subtraction Bingo

| | | | | |
|---|---|---|---|---|
| 80 | 45 | 5 | 44 | 1 |
| 42 | 3/8 | Difference | 54 | 2 |
| Digit | 60 | | Subtraction | 9 |
| 15 | Hundreds Place | 1/9 | 19 | 14 |
| Zero | 7 | Subtrahend | Minuend | 70 |

© Barbara M. Peller

# Subtraction Bingo

| 8 | Borrow | 10 | Digit | 6 |
|---|---|---|---|---|
| 19 | Subtraction | 80 | Minuend | 45 |
| 33 | 30 | | – | 5 |
| 4 | 20 | 90 | 50 | 11 |
| 14 | 16 | 3 | 1/9 | 9 |

Subtraction Bingo: Card No. 9

© Barbara M. Peller

# Subtraction Bingo

| | | | | |
|---|---|---|---|---|
| 1/9 | 44 | 1 | Difference | Digit |
| 54 | 2 | 24 | – | Subtraction |
| 60 | 45 | | 13 | Addition |
| 90 | 36 | 19 | 16 | 33 |
| 3 | 70 | 40 | 20 | 80 |

Subtraction Bingo: Card No. 10

© Barbara M. Peller

# Subtraction Bingo

| | | | | |
|---|---|---|---|---|
| = | 45 | Minuend | 19 | 70 |
| 5 | 33 | 50 | 8 | 24 |
| 42 | Subtraction | | 40 | 10 |
| 3 | 1/6 | 16 | 6 | 1/9 |
| 4 | 7 | 30 | Subtrahend | Borrow |

© Barbara M. Peller

# Subtraction Bingo

| | | | | |
|---|---|---|---|---|
| Borrow | 9 | 33 | 44 | 8 |
| 10 | Zero | 3/8 | Subtrahend | – |
| 30 | 11 | | 54 | Difference |
| 7 | 14 | Subtraction | 1/9 | 42 |
| 45 | 5 | 60 | 4 | 2 |

© Barbara M. Peller

# Subtraction Bingo

| | | | | |
|---|---|---|---|---|
| 4 | 9 | = | 33 | 54 |
| 3/8 | 5 | Subtraction | 8 | Addition |
| 44 | 2 | | 10 | 11 |
| 80 | 16 | 1 | 60 | 1/9 |
| 7 | 36 | Subtrahend | 30 | 50 |

Subtraction Bingo: Card No.13

© Barbara M. Peller

# Subtraction Bingo

| | | | | |
|---|---|---|---|---|
| 6 | Subtraction | Minuend | 8 | 4 |
| 2 | 30 | 33 | – | 45 |
| 19 | 13 | | 40 | 3 |
| 36 | 16 | 60 | 1 | = |
| 7 | Difference | Addition | 70 | 80 |

Subtraction Bingo: Card No. 14

© Barbara M. Peller

# Subtraction Bingo

| | | | | |
|---|---|---|---|---|
| 50 | 8 | Minuend | Borrow | 44 |
| = | 40 | 24 | 3/8 | 19 |
| 54 | 30 | | 1/6 | 45 |
| 7 | 33 | 5 | 16 | 4 |
| 70 | 14 | Subtrahend | Digit | 10 |

Subtraction Bingo: Card No. 15

© Barbara M. Peller

# Subtraction Bingo

| | | | | |
|---|---|---|---|---|
| 1 | 33 | 5 | Digit | 17 |
| Difference | Addition | 11 | 42 | 13 |
| 4 | 9 | | 54 | 10 |
| Hundreds Place | 2 | 7 | 50 | 1/9 |
| 19 | Tens Place | Subtrahend | 14 | 45 |

Subtraction Bingo: Card No. 16

© Barbara M. Peller

# Subtraction Bingo

| | | | | |
|---|---|---|---|---|
| 3 | Ones Place | 12 | 33 | 6 |
| 50 | 19 | 16 | 13 | 11 |
| 8 | 80 | | Tens Place | 5 |
| 20 | 70 | 1/9 | Minuend | Addition |
| 90 | 4 | Borrow | 44 | 9 |

© Barbara M. Peller

# Subtraction Bingo

| | | | | |
|---|---|---|---|---|
| Digit | 60 | 2 | 19 | Difference |
| 45 | 3 | 90 | 54 | 4 |
| 8 | Addition | | 12 | 3/8 |
| 20 | 24 | 16 | 1/9 | 40 |
| Tens Place | 33 | Minuend | Ones Place | = |

Subtraction Bingo: Card No. 18

© Barbara M. Peller

# Subtraction Bingo

| 54 | = | 33 | 5 | 60 |
|---|---|---|---|---|
| 50 | 44 | 45 | Borrow | 13 |
| Ones Place | 6 | | – | 1/6 |
| 40 | Tens Place | 90 | 14 | 12 |
| 3/8 | 17 | 70 | 80 | Subtrahend |

© Barbara M. Peller

# Subtraction Bingo

| 60 | Ones Place | 44 | 33 | – |
|----|-----------|----|----|----|
| 2 | 10 | 42 | 90 | Difference |
| 9 | 11 | | Hundreds Place | 24 |
| 20 | Zero | 15 | 14 | Tens Place |
| 18 | 80 | 17 | 1/9 | 12 |

© Barbara M. Peller

# Subtraction Bingo

| 50 | = | 42 | 33 | 36 |
|---|---|---|---|---|
| 9 | 12 | 1 | 5 | 30 |
| Addition | 70 | | Ones Place | Minuend |
| 90 | Borrow | Tens Place | 20 | 80 |
| Hundreds Place | 17 | Subtrahend | 3 | 14 |

© Barbara M. Peller

# Subtraction Bingo

| | | | | |
|---|---|---|---|---|
| Digit | 40 | 12 | 3/8 | 4 |
| Difference | 44 | 1/6 | 5 | – |
| 2 | 13 | | 30 | 11 |
| Tens Place | 20 | 14 | 24 | 6 |
| 17 | 3 | Ones Place | Addition | 42 |

© Barbara M. Peller

# Subtraction Bingo

| | | | | |
|---|---|---|---|---|
| 1 | Ones Place | Borrow | 3/8 | Subtrahend |
| = | 60 | 70 | 50 | 24 |
| 40 | 4 | | 15 | 30 |
| Addition | 17 | Tens Place | 3 | 14 |
| 36 | Zero | 80 | 90 | 12 |

© Barbara M. Peller

# Subtraction Bingo

| | | | | |
|---|---|---|---|---|
| 1 | 60 | 6 | Ones Place | 5 |
| 12 | Subtrahend | 42 | Difference | 30 |
| 11 | Digit | | 4 | Addition |
| 36 | 15 | Tens Place | 3 | 9 |
| 18 | Hundreds Place | 17 | 44 | Zero |

© Barbara M. Peller

# Subtraction Bingo

| | | | | |
|---|---|---|---|---|
| Hundreds Place | 42 | Ones Place | Minuend | 12 |
| 24 | 36 | 50 | 1 | – |
| 9 | 5 | | 15 | Tens Place |
| 1/6 | 20 | Zero | 17 | 13 |
| Subtrahend | 6 | 2 | 19 | 18 |

© Barbara M. Peller

# Subtraction Bingo

| | | | | |
|---|---|---|---|---|
| 12 | Ones Place | 40 | Difference | Digit |
| 90 | 44 | 5 | 60 | 1 |
| 36 | 15 | | 13 | Hundreds Place |
| 3 | 3/8 | 20 | 17 | Tens Place |
| 11 | 19 | Minuend | Zero | 18 |

Subtraction Bingo: Card No. 26

© Barbara M. Peller

# Subtraction Bingo

| 40 | 2 | Ones Place | 60 | 10 |
|----|----|----|----|----|
| 36 | 15 | 50 | Tens Place | – |
| 16 | Zero | | 17 | Hundreds Place |
| Digit | = | 42 | 18 | 24 |
| 4 | 13 | 12 | 1/6 | 11 |

© Barbara M. Peller

# Subtraction Bingo

| | | | | |
|---|---|---|---|---|
| 54 | 60 | 1/6 | Ones Place | 1 |
| 10 | 12 | 15 | Difference | 13 |
| Zero | Addition | | 11 | 90 |
| 1/9 | Digit | 70 | 17 | Tens Place |
| 3/8 | 8 | 4 | 18 | 36 |

Subtraction Bingo: Card No. 28

© Barbara M. Peller

# Subtraction Bingo

| 12 | 60 | Digit | 50 | 8 |
|----|----|-------|----|-----|
| 36 | 90 | 42 | 11 | 1/6 |
| 9 | 15 | | – | Ones Place |
| 10 | 20 | Subtraction | 17 | Tens Place |
| 1 | 5 | 18 | = | Zero |

© Barbara M. Peller

# Subtraction Bingo

| | | | | |
|---|---|---|---|---|
| 6 | Ones Place | Difference | 8 | Tens Place |
| 24 | 60 | 40 | 13 | – |
| 36 | 4 | | 11 | 42 |
| 18 | = | 3/8 | 17 | 15 |
| 20 | Borrow | Zero | 12 | 1/6 |

Subtraction Bingo: Card No. 30

© Barbara M. Peller

www.ingramcontent.com/pod-product-compliance
Lightning Source LLC
Chambersburg PA
CBHW051428200326
41520CB00023B/7401